MATRIX MAGIC

by

0. 0. Adeyemo BSc., MSc., MCIWEM, CEng.

In memory of Eunice Adeyemo, a wonderful mother who

procured my education.

Lulu.com 2010

First published 2010

ISBN 978-1-4457-2012-8

Preface

MATRIX MAGIC

By O. O. Adeyemo

With the advent of the proliferation of computers and the institutionalisation of software many aspects of traditional and applied maths are hidden from sight in computer packages churning answers that appear by magic and giving immediate and well-polished results. It is very easy in such circumstances to forget the solid, proven maths machinery that lies behind the achieving of such results. For the mathematician, he may be more interested in the methods of achieving the results than the final application. A full knowledge of how they relate to outcomes contributes to his library of ideas and may be essential for his devising new and advanced solutions.

In this document, I wish by the use of choice examples, initiate the appetite of junior mathematicians and scientists and engineers, showing the practical applications of some of their tools, with the full workings exposed for their elucidation. I hope they will be able to use these tools with more confidence in solutions of their own.

The routines and programmes created and used in this project are available in the directory and sub directories of

"Accessory" found as downloads from my website at www.anglo-african.com.

TABLE OF CONTENTS

LIST OF TABLES

LIST OF FIGURES

Chapter 1

APPLICATIONS OF LINEAR ALGEBRA

1.1 Linear Equations

Typically, linear equations solve problems of n variables using m<=n linear independent equations, if m=n then the values of the n variables are unique.

Example 1.

A buyer at a department store has forgotten the unit costs for women's clothing items provided by a specific supplier, but checking his documentation he finds three invoices detailing items and costs as shown below.

Invoice 1

45 blouses + 12 skirts + 12 trousers =£699.9

Invoice 2

20 blouses + 12 skirts + 8 trousers =£421. 2

Invoice 3

20 blouses + 10 skirts + 5 trousers =£365. 5

What were the unit costs for Blouses, skirts and trousers?

Example 2.

A surveyor knows he can make a first estimate on the value of a property in direct relation to the number of bedrooms, receptions and bathrooms in the property. He does not know what values to use but he knows the first estimates for three properties as given below.

Property 1

2 receptions +5 bedrooms + 2 bathrooms =£735k

Property 2

1 bedroom + 1 bathroom = £120k

Property 3

1 reception + 3 bedrooms+2 bathrooms =£445k

What first estimate would he make for a block of 5 tenements each having 2 bedrooms and a bathroom that his company wishes to purchase?

The solution of such problems is eased by the creation of an n x n+1 matrix that we manipulate to reduce into an upper triangular form. The variables are then solved by back substitution.

The two examples given are simple n=3 dimensional problems that are relatively easy to solve manually.

Example 1 can be represented as:

Table 1

line

1.	$45b$	$+12sk$	$+12t$	$= 699.9$
2.	$20b$	$+12sk$	$+8t$	$= 421.2$
3.	$20b$	$+10sk$	$+5t$	$= 365.5$

Use line 2 to subtract from line 3

Table 2

line

1.	45b	+12sk	+12t	= 699.9
2.	20b	+12sk	+8t	= 421.2
3.-2.	0	-2sk	-3t	= -55.7

To eliminate blouses from line 2 we subtract 20/45 x line 1 from line 2.

Reversing the signs throughout line 3 keeps all the resulting elements positive and the result more elegant.

Table 3

line

1.	45b	+12sk	+12t	= 699.9
2.-(20/45)1.	0	+6.67sk	+2.67t	= 110.13
-3.	0	+2sk	+3t	= 55.7

We eliminate sk in line 3 from by subtracting 2/6.67 x line 2 from line 3

Table 4

line

1.	45b	+12sk	+12t	= 699.9
2.	0	+6.67sk	+2.67t	= 110.13
3.-(2/6.67)2.	0	0	+2.20t	= 22.68

The solution is given by t = 22.68/2.20 = 10.31

And then back substitution:-

6.67 sk =110. 13 – 2.67 x 10.31

sk = 12.38

45 b =699.9 – 123.72 – 148.56

b = 427.62/45 = 9.50

What is observed doing this relatively simple example, is the quantity of calculations with the possibility of error and how rounding errors carry right through the calculation to give slightly erroneous results, thus the price of skirts, sk = 12.38 instead of a true value of 12.40 and trousers t =10.31 instead of a true value of 10.30.

Automating the process would have great value, releasing our need to do tedious calculations and rounding errors can be reduced to a minimum. Using a spreadsheet we can generate an instant result.

Download and open LinEqSoln.xls.

For good programming practice create a copy of this spreadsheet using the 'save as' option and change its name e.g. "Example1.xls". On page Linear Equations, at the head of the page we have a 10 x 11 matrix that represents a series of linear equations to be solved. From table 1 we note we may write he formulation of this problem in reduced form:

45	12	12	=699.9
20	12	8	=421.2
20	10	5	=365.5

If we drop the equal signs fill it with leading zeros and make it the last three lines of our opening matrix (make sure you only adjust the lead matrix) as in fig. 1.

Fig 1.

		Microsoft Excel - chptr1.xls												
		A	B	C	D	E	F	G	H	I	J	K	L	M	N	O

Matrix Manipulations

A x X = Y

C	D	E	F	G	H	I	J	K	L	N
1	2	3	4	5	6	7	8	9	10	1
20	19	18	17	16	15	14	13	12	11	3
21	1	2	3	4	5	6	7	29	30	2
31	32	1	2	3	4	5	6	7	40	3
41	31	32	1	2	3	4	5	6	7	4
51	31	32	54	1	2	3	4	5	6	6
61	62	63	16	15	19	18	17	16	70	3
0	0	0	0	0	0	0	45	12	12	699.9
0	0	0	0	0	0	0	20	12	8	421.2
0	0	0	0	0	0	0	20	10	5	365.5

We find the 9 following matrices use a systematic Gauss elimination to produce an upper triangular matrix. The 9th matrix down holds in the bottom right hand corner a 3 x 4 upper triangular solution of our linear equations. Shown in fig 2.

Fig 2.

1	2	3	4	5	6	7	8	9	10		1
0	-21	-42	-63	-84	-105	-126	-147	-168	-189		-17
0	0	21	42	63	84	105	126	168	189		14.19048
0	0	0	32	64	96	128	160	255	319		19.66667
0	0	0	0	32	64	96	128	253	317		18.84354
0	0	0	0	0	54	108	162	422.625	530.625		32.43516
0	0	0	0	0	0	-5	-10	-12.31944	34.68055511		-6.56136
0	0	0	0	0	0	0	45	12	12		699.9
0	0	0	0	0	0	0	0	6.666667	2.666666667		110.1333
								0	-2.2		-22.66

13

The 10th matrix down shows the back substituted solutions our problem, beginning with the last variable and progressing, Shown in fig 3. Thus trousers have a unit cost £10.30

skirts have a unit cost £12.40

blouses have a unit cost £9.50

Fig 3.

X	DET	Dimension 1x1	2x2	3x3	4x4	5x5	6x6	7x7	8x8	9x9	10x10
		1	-21	-441	-14112	-451584	-24385536	2.44E+08	10973491387	7.32E+10	-1.6E+11
10		0	0	0	0	0	0	0	0	0	10.3
9		0	0	0	0	0	0	0	0	16.52	12.4
8		0	0	0	0	0	0	0	15.55333333	11.148	9.5
7		0	0	0	0	0	0	-6.936111	-38.04277779	-69.9356	-70.0047
6		0	0	0	0	0	0.600651	14.47287	30.02620671	-22.2642	-86.1489
5		0	0	0	0	0.588861	-0.612442	-7.548553	-7.548552856	79.72064	144.829
4		0	0	0	0.614583	-0.563138	0.037513	0.037513	0.037513386	0.324319	0.431611
3		0	0	0.675737	-0.55343	0.035431	0.035431	0.035431	202.2287642	145.4757	134.5448
2		0	0.809524	-0.54195	0.072633	0.072633	0.072633	0.072633	-404.3140334	-289.259	-256.518
1		1	-0.61905	0.056689	0.056689	0.056689	-0.056689	0.056689	202.2500227	128.4607	2150.557

Examine example 2 and we can rationalise the problem into table 5:

Table 5.

bedroom	reception	bathroom	Value £ 000
5	2	2	=735
1	0	1	=120
3	1	2	=445

We note the second line has a zero for the second element. This could cause problems in the elimination process if it lies in the leading diagonal of our ten line matrix. In such case we would need to manipulate the zero to an alternative part of our matrix, preferably below the leading diagonal. We can manipulate the matrix without changing its intrinsic value by changing the position of a complete row or a complete column. If we change the position of a column we have to remember we are changing the order of variables. The reduced solution looks like:

1	0	1	= 120
5	2	2	= 735
3	1	2	= 445

Put this into the bottom right hand side of the opening matrix and filling the left hand cells with zeros.

The results can immediately be seen in matrix 10

Bathroom =35

Reception =120

Bedroom =85

It follows the preliminary estimate for a block of 5 tenements each with two bedrooms and a bathroom based on these values would be

10 x 85 + 5 x 35 = £1025k.

Example 3

Having shown some commercial uses of linear equations, we may now go on to see its use in electric circuit analysis. Consider the circuit shown below in fig 4. What would be the potential at A with the switch closed or alternatively with the switch open?

Fig4.

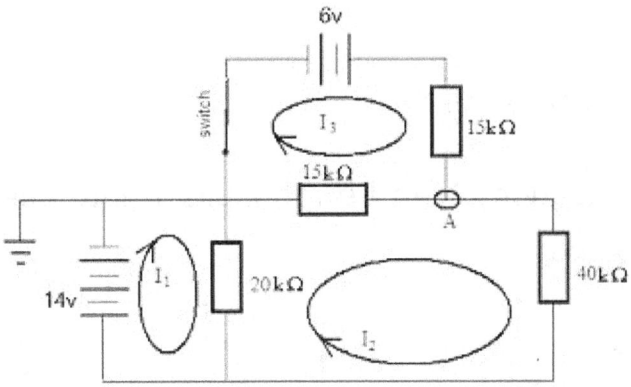

The secret of answering this problem is in the choice of a minimum number of loop currents that pass through all the major elements of our circuit. We then use Kirchhoff's second law for each loop, i.e. the sum of the voltage drops across each element of the loop equates to the potential difference driving the loop, in other words the sum of the potentials in going round any loop of the circuit from a starting point back to itself is zero.

The current circuit has been analysed into 3 loops, I_1, I_2 and I_3 as seen.

Using Ohms Law, V=IR on the three loops we show,

Loop1:

$20(I_1-I_2)= -14$

Loop 2:

$20(I_2-I_1) + 15(I_2-I_3) + 40I_2 = 0$

Loop3

$15I_3 + 15(I_3-I_2) = +6$

these equations can be rationalised into linear equations of I_1, I_2 and I_3.

Table 6.

I_1	I_2	I_3	V
20	-20	0	= -14
-20	+75	-15	= 0
0	-15	+30	= +6

Entering this matrix in the appropriate fashion into our spreadsheet gives solutions $I_3 = 0.084211$ milliamps, $I_2 = -0.23158$ milliamps and $I_1 = -0.93158$ milliamps.

If we assume earth to be at 0v it follows that A will be at $-(I_2 - I_3)*15k$

$= 4.73684v$.

It can be shown that the alternative route to A through the 6V battery will give $+6 - I_3*15k = 4.73684v$, this is a confirmation of our answer.

18

When the switch is thrown open, I_3 ceases, observing loop 2 we note

$(15 + 40)k * I_2 = -14$, $I_2 = -0.24546$ milliamps. The potential at A is then

- $(-0.24546*15)(k*milliamps) = 3.6819v$.

Now we may consider a more elaborate example that shows how to analyse a more complex circuit.

Example 4

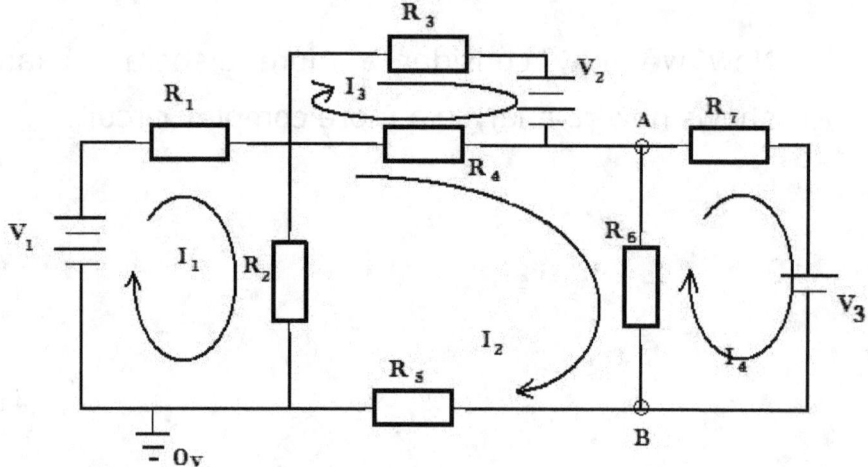

Fig.5

Consider Fig 5 above. Use method of solution of linear equations to solve the following questions given :

V1 = 24v, V2 = 8v, V3 = 3v

R1 = 4Ω, R2 = 20Ω, R3 = 5Ω, R4 = 4Ω, R5 = 6Ω, R6 = 50Ω,

R7 = 10Ω.

a) Find the voltage across load R6.

20

b) R2 and R7 have precision ±10%, find the affective range across R6 attributable to this variation.

c) If V2 is reversed, what is the potential at A.

Solution:

The circuit has already been analysed into 4 loop currents I1 to I4.

The analysis of these loops follow.

Loop 1

I1*R1 + (I1-I2)R2 = V1

Loop 2

(I2-I1)R2 + (I2-I3)R4 + (I2-I4)R6 + I2*R5 = 0.0

Loop 3

I3*R3 + (I3-I2)R4 = -V2

Loop 4

I4*R7 + (I4-I2)R6 = -V3

These equations can be written in reduced form in Table 7 below

Table 7.

I1	I2	I3	I4	
R1+R2	-R2			=V1
-R2	R2+R4+R6+R5	-R4	-R6	=0
	-R4	R3+R4		=-V2
	-R6		R7+R6	=-V3

In figures this becomes

I1	I2	I3	I4	
24	-20	0.0	0.0	=24
-20	80	-4	-50	=0.0
0.0	-4	9	0.0	=-8
0.0	-50	0.0	60	=-3

To solve this we can enter the values and use the process detailed in our use of linEqSoln.xls page Linear Equations. Alternatively, a possibly more convenient solution method is to run the application program Gaussol.exe found in folder 'Accesssory'. Prior to running this program the input file "LinearIn.txt" has to be updated with the matrix representing the linear equation to be solved.

In this particular case the entry would be as follows:-

4

24   -20   0.0   0.0

-20   80   -4   -50

0.0   -4   9   0.0

0.0   -50   0.0   60

24

0

-8

-3

This must be saved in the same directory and then we can execute Gaussol.exe. A check of details of output file LinearOut.txt should now show that it has been freshly created. Open this file to obtain the solution vector in the last n lines of the file.

$I1 = 1.528$
$I2 = 0.715$
$I3 = -0.571$
$I4 = 0.546$

From examination of the circuit we can establish potentials for A and B.

The potential at A = (I1-I2)R2 +(I3-I2)R4 =12.549v.

The potential at B = I2*R5 or A-I4*R7-3v =4.207v.

a) The voltage across load R6 is A-B =8.342v

b) The voltage across R6 is independent of R2 (above we show B= A-I4*R7-3v), as such we investigate new solutions with R7=9Ω and 11Ω .

R7=9Ω, A=12.418v B=4.348v A-B=8.07v

R7=11Ω, A=12.668v B=4.079 A-B=8.589

c) We can reverse V2 by making it V2=-8v

Line 3 of our linear equation then becomes

0.0   -4   9   0.0 =+8.

When we run the solution we obtain:

I1 =1.882
I2 =1.058
I3 =1.359
I4 =0.832

Which in turn solves to A=17.674v

This type of solution can be utilised in solving the more general AC circuits also. To do this we must consider the voltage to be of the form V=V(cosωt + Jsinωt) and impedance Z =R+J(ωL-1/ωC).

Example 5.

Fig 6.

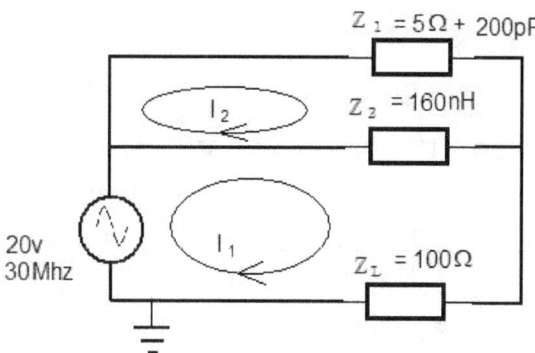

a) Find the load current across ZL in figure 6 above.

b) Find the magnitude and phase of the voltage across ZL if the input frequency is increased to 50 Mhz.

This is an AC circuit powered by a 30Mhz 20volt supply and a 100 ohm load, it has been analysed into 2 loops.

The reactance of Z1 is calculated by $Zc = -j/\omega C$, $\omega = 2\pi f$ = 1.88E+8. $Zc = -j26.5$.

The reactance of Z2 is $j\omega L = +j30.2$.

Loop 1

$(I1-I2)Z2 + ZL*I1 = 20$

Loop 2

$I2*Z1 + (I2-I1)*Z2 = 0$

These form the linear equations:

Z2+ZL	-Z2	= 20
-Z2	Z1+Z2	=0

Substitute in the values:

100+j30.2	0.0 -j30.2	=20+j0.0
0-j30.2	5+j3.7	=0.0+j0.0

These are the complex linear equations to be solved. In directory 'Accessory' application CGaussol.exe is used for solving linear equations in complex numbers. Make the correct input into CLinSolIn.txt and run the program. The solution is given in ClinSolOut..txt..

a)

I1 = 0.0859 +j 0.0225

I2 = 0.160   +j 0.400

We can verify this, if we use Thevinin's theorem of an equivalent circuit. Thus with ZL removed the total circuit admittance = 1/Z1+1/Z2.

$$= \frac{1}{5-j25} + \frac{1}{0+j30.2} = \quad 0.006862 + j\,0.003248 \quad .$$

When we convert this to impedance and add ZL, the total impedance of the circuit is   219.05 - j 56.35. Current I1 is obtained from I=V/Z

$$I1 = \frac{20 + j0}{219.05 \ - j\,56.35} = 0.0856 + j\,0.0220 \; .$$

c) Increasing input frequency to 50 Mhz, the reactance of Z1 = -j15.9 and the reactance of Z2 changes to j50.3. The Linear equations

Z2+ZL   -Z2   = 20

-Z2   Z1+Z2 =0

become

100 + j 50.3   0.0 - j50.3   = 20.0 +j 0.0

0.0 - j50.3   5 – j 34.3   = 0.0   + j0.0

27

with solution

I1 = 0.174 + j 0.0345

I2 = 0.243 + j 0.0860

The potential across ZL is ZL*I1 = (100.0 + j 0.0) I1 = 17.4 + j 3.43

With magnitude = 17.7v

and phase angle $\tan^{-1}\left(\dfrac{3.43}{17.4}\right) = 11.15°$

1.2 Linear Programming.

Example 6

Consider the problems faced by a factory production manager. His factory produces a range of 4 mobile phones, products P1, P2, P3 and P4 produced on 5 machines M1, M2, M3, M4 and M5.

For each unit produced of product P1, he must use M1 for 3 hours, M3 for 5 hours, M4 for 2 hours and M5 for 2 hours.

For each unit produced of product P2, he must use M1 for 2 hours, M2 for 2 hours, M3 for 3 hours and M5 for 2 hours.

For each unit produced of product P3, he must use M1 for 2 hours, M2 for 2 hours, M4 for 2 hours and M5 for 2 hours.

For each unit produced of product P4, he must use M1 for 2 hours, M2 for 1 hour, M3 for 1 hours, M4 for 1 hours and M5 for 1 hours.

The respective profits per unit produced are P1 £950, P2 £350, P3 £450 and P4 £200.

In any 1 week M1 runs for a maximum 192hours, M2 192 hours, M3 96 hours and M5 192 hours

a) If he can sell all he produces, how much of each product would he produce and what would be his profit?

b) If he has a confirmed order for 9 units of P1 that he must meet, how would he maximize profit on his remaining resources and what would be his product run?

c) If he is allowed to covert 48 hours on any machine to 48 hours on another what exchange would he make to maximise his profits under constraints given in b).

To solve this problem, we return to spreadsheet LinEqSoln.xls, go to tab Lin Programming, you will find the page set up for the solution. To ease explanation I will refer to cells on the page by their cell references.

Row 7 lists the various production machines, column C lists the products by number. Row 8 lists the maximum working hours for each machine. Each product uses 2 rows, thus product 1 uses rows 12 & 13, product 2 uses rows 14 & 15 etc. Column D shows the entered value of nominal profit per item produced, the figure in column D second product cell enclosed with a bold border is the number of products we propose to produce. The first row of each product details the entered value of the number of hours per machine in the production of the product, thus cell G5 shows we require 5 hours on machine 3 to produce a unit of product 1. The second row of the product indicates the number of hours used on each machine to produce the proposed number of the product and column K shows the profit of producing that number of products. Cell K22 shows the total profit across all the products. The values of hours per machine per unit product, and profit for each product are user adjustable and so the table is easily adaptable for another problem.

The way the table has been set originally is to propose just 1 unit of each product. Looking at product 1 we see, as expected we have consumed 3 hours from M1, 5 hours from M3, 2 hours from M4, 2 hours from M5 and produced profit of £950. The total profit is £1950. Change cell D13 to 2, i.e. we want to produce 2 units of P1. We can immediately see in row 9 how the available hours for each machine are decreased, how the profit from P1 is increased to £1900 and how the total profit has been incremented to £2900.

This is obviously not a solution of our problem but it shows how the table works and how we can use it to produce solutions.

a) Product P1 has the largest profit per unit, if we want to produce this to a maximum it will be limited by machine 3. Try producing 19 units of P1. This uses 95 hours of M3 and leaves only enough time for 1 unit of P4, giving a total profit of £18250.

Product P1 seems to be consuming too many hours of M3 to give an optimum result, let us try 17 P1. The reduction in 2 units of P1 releases 10 hours on M3, this can be used by 5 units of P3. the resulting profit has increased to £18600.

P4 uses only 1 hour of M3 per unit, let us try producing 96 units. This produces a profit of £19200.

If we prioritise the production of P3 we can produce 48 units with a profit of £21600. This appears to be the maximum profit we can gain if we are unrestricted in production.

b) The required 9 units of P1 utilize 45 hours of M3, leaving just 51 hours for further use, this being the limiting constraint on further production. It is easily confirmed that under the constraints, 9 P1, 25 P3 and 1 P4 is the optimum giving a profit of £20000.

c) The results from b) show a residual 141 hours available on M2. Set the maximum hours available on M2 to144 and increase the maximum available hours for M3 to 144. Under this regime, the 9 units of P1 use 45 hours on M3 leaving 99 hours. If we try to produce 45 units of P3, M4 becomes 20 hours in deficit, showing it is not feasible. We can only produce 39 units of P3. The remaining 21 hours of M3 to produce 7 units of P2. The solution is then 9 P1, 7 P2, 39 P3 and the profit is then increased to £28550.

1.2b Classical Transport

The classical transport problem is the solution of how to supply a number of consumer points with a specific merchandise supplied from a number of suppliers, the unit price of merchandise from each supplier is dependent on which consumer he supplies, in effect the price differentiation may depend on distance and transportation or administrative or storage costs. The network of consumer points may represent an organisation that is trying to minimise its supplier's costs.

Here I demonstrate with a simple example how we can solve this with a spreadsheet solution, as shown in LinEqSoln.xls.

Example 7

We have 3 warehouses a,b,c ready to supply 3 retailers A,B,C. The matrix of costs of suppliers against consumers for a given item is shown below.

Supplier		a	b	c
Destinations	A	6	10	15
	B	4	6	16
	C	12	5	8

Initial stocks are such that a holds 1 item, b holds 8 items and c holds 7 items. The problem is to work out the minimum costs to deliver 2 items to A 5 items to B and 9 items to C.

Open spreadsheet LinEqSoln and go to page 'transport'.

In line 10, warehouse 1 represents a, 2 represents b and 3 represents c. In column F destination 1 represents retailer A, 2 represents B and 3 represents C.

In line 11 we can insert a total stock for a as 1, b as 8 and c as 7 and all others as 0.

In line 14 we enter costs per item delivered retailer A, similarly lines 17 and 20 represent costs for supplying B and C respectively.

For a first solution we may try 1*aB + 2*bA + 4*bB + 2*bC + 7*cC giving subtotals 4 + 20 + 24 +10 + 63= 121.

A better solution is found as

1*aA + 1*bA + 5*bB + 2*bC +7cC giving subtotals

6 + 10 + 30 + 10 + 63= 119.

The most efficient solution is found by trial and error by filling in elements of the table.

There are a few precautions we must observe while using this table, to ensure we obtain a credible and correct result. As the table is designed to be used for up to 10 warehouses and 9 destinations, it is important that all entries for non existent warehouses and destinations are zeroed.

This table is used when there is a static cost per item delivered dependent on its source warehouse. When there are volume discounts or bulk freight charges it would have to be revised. For example in this example we require 2 items to be delivered to retailer A. The cheapest option appears to be 1 item from a and a second item from b at a total cost of 6+10=14 cost units. In practice, sending a vehicle from a to A and b to A may in effect be doubling of transport costs and it might be cheaper and more practical to just get the 2 items required from warehouse b.

Chapter 2

PROJECTILE WITH DRAG

In the ideal with no friction loss, a projectile has velocity and acceleration that can be resolved in the XY plane and the Z direction. In the XY plane we assume after release the projectile has no force acting on it and acceleration is zero. In the Z direction, after release the projectile suffers a deceleration of –g that causes it to eventually fall back to ground. This is described in typical introductory physics. The problem becomes more realistic if we account for frictional losses due to air drag.

In the XY plane, assume a deceleration proportional to the square of velocity, $a = \dfrac{F}{m} = -cu^2$ where c is a constant.

$$\frac{du}{dt} = -cu^2 \qquad eqn\ 2.01$$

integrate to

$$\int_{t_0}^{t_f} dt = -\int_{u_0}^{u_f} \frac{du}{cu^2} = \frac{1}{c}\left[\frac{1}{u}\right]_{u_0}^{u_f} \quad eqn\ 2.02 \ .$$

In the Z direction on the way up (u +ve), we augment gravity with a deceleration proportional to the square of the velocity.

36

$$a_z = \frac{F_z}{m} = -g - cu^2 = \frac{du}{dt} \qquad eqn\ 2.03$$

Forms the integration

$$-\int dt = \int \frac{du}{g + cu^2} \qquad eqn\ 2.04$$

To solve this make the substitution

$$cu^2 = g \tan^2 \theta \qquad eqn\ 2.05$$

$$u = \sqrt{\frac{g}{c}} \tan \theta \qquad eqn\ 2.06, \quad \theta = \tan^{-1}\left(\sqrt{\frac{c}{g}}\, u\right) \qquad eqn\ 2.07$$

$$du = \sqrt{\frac{g}{c}} \sec^2 \theta\ d\theta \qquad eqn\ 2.08$$

Eqn 2.04 becomes

$$-\int dt = \frac{1}{g} \int \sqrt{\frac{g}{c}} \frac{\sec^2 \theta}{\sec^2 \theta}\ d\theta$$

$$= \sqrt{\frac{1}{gc}} \int_{\theta_0}^{\theta_f} 1\, d\theta\ ,\quad \theta_0 = \tan^{-1}\left(\sqrt{\frac{c}{g}}\, u_0\right),\ \theta_f = \tan^{-1}\left(\sqrt{\frac{c}{g}}\, u_f\right)$$

$$= \sqrt{\frac{1}{gc}}\ [\theta]_{\theta_0}^{\theta_f}$$

$$= \sqrt{\frac{1}{gc}}\left(\tan^{-1}\left(\sqrt{\frac{c}{g}}\, u_f\right) - \tan^{-1}\left(\sqrt{\frac{c}{g}}\, u_0\right)\right) \qquad eqn\ 2.09$$

For Z direction on the way down (u –ve) acceleration is described,

$$a_z = \frac{F_z}{m} = -g + cu^2 = \frac{du}{dt} \qquad eqn\ 2.10$$

$$-\int dt = \int \frac{du}{g - cu^2} \qquad eqn\ 2.11$$

Let $cu^2 = g\cos^2\theta \qquad eqn\ 2.12$

$$u = \sqrt{\frac{g}{c}}\cos\theta \qquad eqn\ 2.13,$$

$$\theta = \cos^{-1}\left(\sqrt{\frac{c}{g}}u\right) \qquad eqn\ 2.14$$

$$du = -\sqrt{\frac{g}{c}}\sin\theta\,d\theta \qquad eqn.2.15$$

$$\int \frac{du}{g - cu^2} = -\sqrt{\frac{g}{c}}\int \frac{\sin\theta}{g\sin^2\theta}d\theta = -\frac{1}{\sqrt{gc}}\int \frac{1}{\sin\theta}d\theta \qquad eqn\ 2.16$$

Consider $I = \int \frac{1}{\sin\theta}d\theta = \int \frac{\sin\theta}{\sin^2\theta}d\theta$

Integrate by parts $\int u\,dv = [uv] - \int v\,du \qquad eqn\ 2.17$

$$\int \frac{\sin\theta}{\sin^2\theta}d\theta = \left[\frac{-\cos\theta}{\sin^2\theta}\right] - 2\int \frac{\cos\theta\cos\theta}{\sin^3\theta}d\theta \qquad eqn\ 2.18$$

Consider $\int \frac{\cos\theta\cos\theta}{\sin^3\theta}d\theta$, integrate by parts

$$\int \frac{\cos\theta\cos\theta}{\sin^3\theta}d\theta = \left[\frac{\sin\theta\cos\theta}{\sin^3\theta}\right] + \int \frac{\sin\theta\sin\theta}{\sin^3\theta} +3\frac{\sin\theta\cos\theta}{\sin^4\theta}d\theta$$

$$= \left[\frac{\cos\theta}{\sin^2\theta}\right]+\int \frac{1}{\sin\theta}+3\frac{\cos\theta}{\sin^3\theta}\,d\theta \; eqn\,2.19$$

$$= \left[\frac{\cos\theta}{\sin^2\theta}\right]+I-\frac{3}{2}\left[\frac{1}{\sin^2\theta}\right] eqn\,2.20$$

substituting back into equation 2.18

$$I = \left[\frac{-\cos\theta}{\sin^2\theta}\right]-2I-2\left[\frac{\cos\theta}{\sin^2\theta}\right]+3\left[\frac{1}{\sin^2\theta}\right] eqn\,2.21$$

$$3I = -3\left[\frac{\cos\theta}{\sin^2\theta}\right]+3\left[\frac{1}{\sin^2\theta}\right]$$

$$I = -\left[\frac{\cos\theta}{\sin^2\theta}\right]+\left[\frac{1}{\sin^2\theta}\right] eqn\,2.22$$

Use equation 2.14

$$\cos\theta = \sqrt{\frac{c}{g}}u,\; \cos^2\theta =\frac{c}{g}u^2$$
$$\sin^2\theta = 1-\frac{c}{g}u^2$$

We may now write equation 2.16

$$\int \frac{du}{g-cu^2}=-\frac{1}{\sqrt{gc}}\int \frac{1}{\sin\theta}\,d\theta =-\frac{1}{\sqrt{gc}}\left(\frac{1}{1-\frac{c}{g}u^2}\left(1-\sqrt{\frac{c}{g}}u\right)\right) eqn\,2.23$$

Example 8

Consider a football with a diameter 0.21m and mass 0.525kg and c=0.0075.

The coefficient c depends on a number of factors, the type of material the ball surface is made from, the roughness of its surface, the air density that in turn will depend on weather, altitude and humidity conditions.

If the ball is kicked with a maximum velocity of 18m/s at 45° to the horizontal, find the maximum distance it will travel and compare it with the frictionless case.

In the directory "Accessory/Projectile" you can find the shortcut to trajectory.exe. This file uses the results obtained in equations 2.02, 2.09 and 2.23 to calculate the trajectories of an object suffering a drag force proportional to the square of the velocity given initial velocity and drag factor c. The resultant table is plotted into a visible chart in trajectory.xls.

To tackle the problem at hand we need to work out components of initial velocity vertical and horizontal.

Vertically in the z direction $V0z=18 \sin45°= 12.73$ m/s.

Horizontally, V0xy=18 cos45°= 12.73 m/s.

Run trajectory.exe by activating the shortcut, a dialogue box should appear on screen. The input boxes on the dialogue box allow you to enter values for c and initial and final velocities in the vertical and horizontal directions. We require to set c to 0.0075, initial vertical velocity to 12.73m/s and initial horizontal velocity to 12.73m/s. We may set final vertical velocity to –12m/s and we may set final horizontal velocity to 5m/s. Press button "Go!" in working directory Accessory/Projectile you find a freshly created text file TrajOut.txt, this file contains the results. Open trajectory.xls and go to page Data and set the selected cell to A8. From the menu bar select "Data/Refresh Data", locate TrajOut.txt and enter it.

To the right of the page are 4 charts that compare velocity and distance travelled for a projectile with air resistance and without air resistance. The top 2 charts show the vertical direction and the lower 2 charts show the horizontal direction. In the upper half of the data page column D represents the vertical level of the projectile at time t, column F shows the vertical level at time t without air resistance. In column D we find z=0 in the period between 2.53s and 2.69s in both cases. With air resistance we may interpolate the time the projectile hits the ground as 2.65s from the horizontal distance table, column D we can estimate the distance travelled as 30.1m. Without air resistance we expect the projectile to hit the ground at

2.59s. The horizontal distance travelled in that time will be 33.0m. Thus the difference between the distance travelled in the two conditions is about 3m or the air resistance reduces the flight by about 10%. Variation in initial conditions can be tested and displayed changing entry values in trajectory.exe and re-importing the resultant output data file on page "Data" of trajectory.xls.

In conclusion we find the velocity of the ball is affected mildly by air friction and the cumulative effect is to shorten its flight by about 10%.

Example 9.

Fig 7.

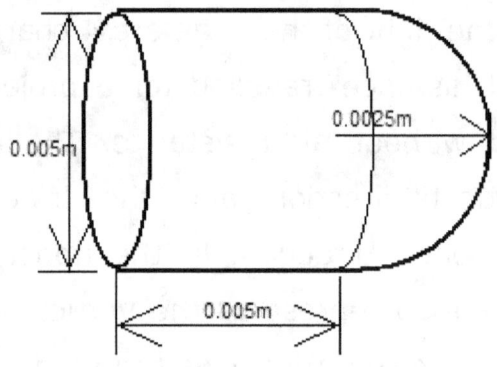

steel bullet, roughness e=46 μm, ρ=7860 kg/m^3

A steel bullet consisting of a hemisphere attached to a cylinder as illustrated in figure 7 above is fired with a velocity of 300m/s at 60° to the horizontal, find its deviation from the path calculated without air resistance at a horizontal distance of 100m away from firing.

The first task is to estimate values for c. We estimate Reynolds number $Re = \frac{\rho V d}{\mu}$, ρ and μ in this case refer to the fluid. For air we may assume $\rho = 1.19 \, kg/m^3$ and μ $= 1.83*10^{-5}$ Ns/m^2. Diameter d is a characteristic length in the object, which we can take as 0.005m. Putting these values in, Re = 97500. From a reference data book, we use the Moody diagram with this value of Re and e/D value 0.000046/0.005 to estimate f = 0.0375.

From f we can estimate shear stress on the surface $\tau = f \frac{\rho}{2} u^2$.

This represents the shear stress on the cylindrical wall. We may assume a C_d Factor of 0.42 for the hemispherical bullet head.

The retarding force on the bullet is made of two components.

$$F = \frac{\rho u^2}{2}\left(\pi D^2 f + \frac{\pi D^2}{2}C_d\right) = u^2 \cdot \frac{\rho \pi D^2}{2}\left(f + \frac{C_d}{2}\right) \qquad Eqn. \ 2.24$$

Force applied F= 300^2 x

$$\frac{1.19\pi 2.5\times10^{-5}}{2}(0.0375 + 0.21) = 300^2 \times 1.16\times10^{-5} = 1.04N.$$

Acceleration $a = \frac{F}{m}$.

Mass $m = \rho\left(\frac{2}{3}\Pi r^3 + 0.005\Pi r^2\right) = \frac{8}{3}\rho\Pi r^3 = 0.001029kg$

We the obtain c = $\frac{a}{u^2}$ =0.0112.

The initial velocity in z direction is 259.81m/s, the initial velocity horizontally is 150m/s. Run trajectory.exe and enter the correct initial conditions. Import the resultant data to page "Data" of trajectory.xls.

We may interpolate from the table of horizontal distance a time to travel 100m with air friction of 1.23s and at that time has a vertical height of 132m. Without air resistance column K shows it will have a horizontal distance of 100m

43

at 0.667s at this time the height under the condition of no air resistance would be 171m. Thus the air resistance causes the projectile to be delayed by 0.56s with a difference in altitude of 39m. We conclude the air resistance considerably alters the trajectory and reduces the speed of the projectile.

TRANSFORMATIONS

In this section we see how a lattice of 3 dimensional position vectors can be manipulated using matrix manipulations. This has advantage when making symmetrically similar replications of an object and when making a realistic representation of the 3 dimensional object in planar 2 dimensional space.

We consider the symmetric transformations of a mirror reflection in a general plane and a rotation of an object about a general axis. If we can generalise these cases for arbitrary objects, mirror surfaces and axes of rotation. We will then posses a very powerful tool.

For the purposes of this section, I have created latticework objects characterised by a series of points held in text files. These are Arrow.txt, Dart.txt and House.txt and are found in directory Accessory/Transform.

At the foundation of our vector manipulations are the basic functions of :
    a) Vector scalar product.
    b) Vector cross product.
    c) Frobenius norm of a vector.

Since we are dealing with 3 dimensional space only, we restrict ourselves to deal with 3dimensional vectors and thus we can produce standard formulas of these functions. Suppose we have vectors Va and Vb, if in 3 dimensions a vector V is composed of 3 components that may be described as Vx, Vy an Vz.

The scalar product of Va and Vb, written Va.Vb is Vax*Vbx + Vay*Vby + Vaz*Vbz.

The cross product results in a vector that is perpendicular to both initial vectors. Va × Vb in terms of Vax, Vay, Vaz and Vbx, Vby, Vbz is described as $\begin{vmatrix} 1 & 1 & 1 \\ Vax & Vay & Vaz \\ Vbx & Vby & Vbz \end{vmatrix}$, the determinant of the matrix. In 3 dimensions

Va × Vb = $\hat{i}$ (Vay.Vbz –Vaz.Vby)+ $\hat{j}$ (Vaz.Vbx –Vax.Vbz) + $\hat{k}$ (Vax.Vby –Vay.Vbx), where $\hat{i}$, $\hat{j}$, $\hat{k}$ are the unit vectors in x, y, z directions.

The Frobenius norm is the magnitude of a vector

$$Fn(Va)= \sqrt{V_{ax}^2 + V_{ay}^2 + V_{az}^2}$$

DotCross.dll is a compilation library of functions that perform these calculations. Programs that are compiled with this library file will have these functions available.

To define a general plane we need a point p with position vector $\begin{pmatrix} Px \\ Py \\ Pz \end{pmatrix}$ on which the plane lies and two non-parallel vectors V1 and V2 which define the direction of extension in the plane.

For a mirror transformation of point X with position vector $\begin{pmatrix} x \\ y \\ z \end{pmatrix}$ we to find the perpendicular distance between the point X and the mirror plane.

Fig.8

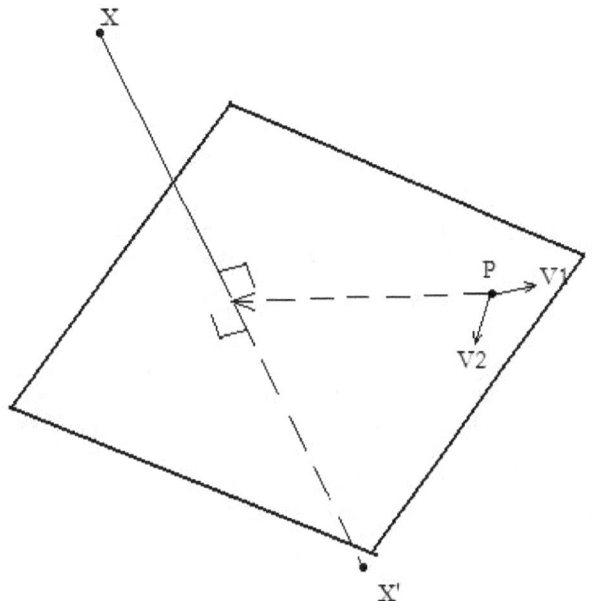

We do this using the vector displacing P onto X. P->X = X-
P we may call this vector Vpx.

Fig.9

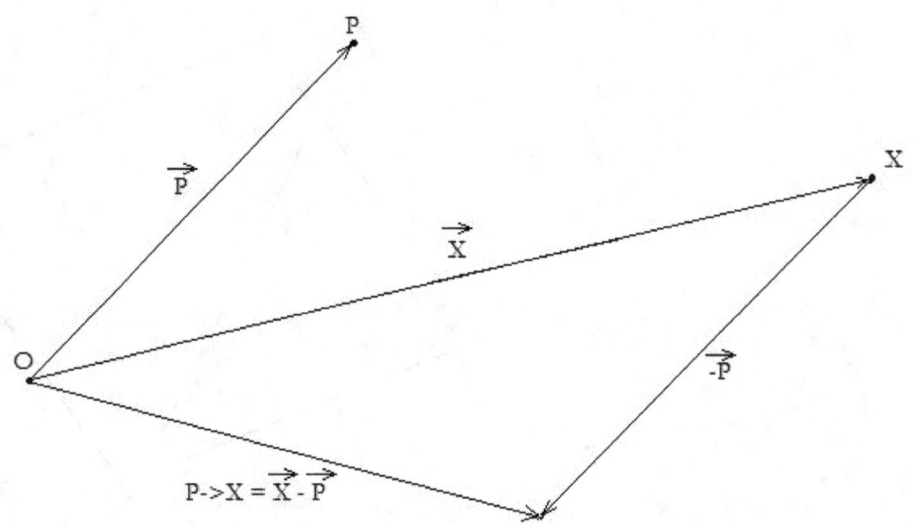

Vpx has a component perpendicular to the mirror surface and a component parallel to the mirror. V1 × V2 gives a vector perpendicular to the mirror, we may normalize this by dividing by the Frobenius norm of the vector. This is a unique vector normal to the plane of the mirror of unit length, we may call this vector V1crV2.

We can now reduce Vpx into Vpx.V1crV2 and Vpx – Vpx.V1crV2. these now our two components perpendicular and parallel to the mirror.

To find the transformed position of our point we add to the position vector of P to the parallel component of Vpx and subtract the perpendicular component.

These actions are realised and used to transform a lattice of points into their reflections within the program mirror.exe, which is a C# developed in C# Microsoft Visual Studio.net 2003 environment. The source program and resources are available in the directory Accessory/Transform.

To define a rotation we need an axis of rotation and a point through which the axis runs. The point P having position vector $\begin{pmatrix} Px \\ Py \\ Pz \end{pmatrix}$ and the axis is a directional vector V with components $\begin{pmatrix} Vx \\ Vy \\ Vz \end{pmatrix}$. Given the axis, we then need an angle of rotation that should ideally be between 0° and 360°.

Fig 10.

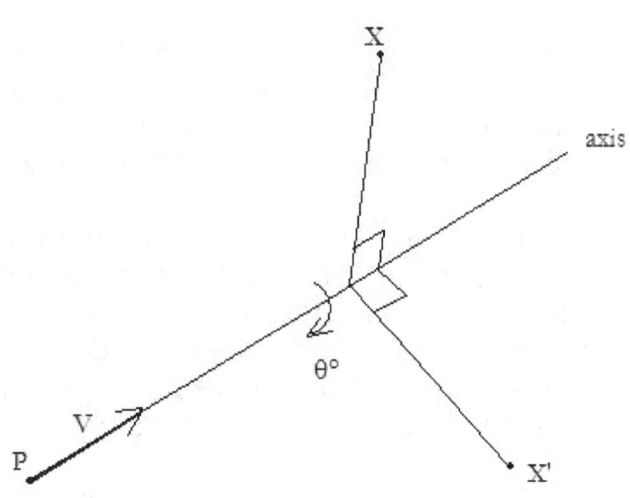

Again, we need to obtain the perpendicular distance between X and the axis of rotation. Again we use the vector P ->X obtained from the position vectors P and X and can be called Vpx. The component of Vpx parallel to the axis of rotation is Vpx .V and the component perpendicular

49

is Vpx –Vpx.V. The perpendicular component can be norm-alised by dividing it by the Frobenius norm and it then be-comes a unit vector, which we may call Vperp. We may use the convention of right hand system of axes to work out a cross product between this perpendicular and the axis vec-tor. V2 = Vperp×V, this vector is perpendicular to the axis and at right angles to Vperp, it will be used in determining the new position of X.

The new position of X under rotation θ, is then given as P + parallel component of Vpx + $\cos\theta$Vperp + $\sin\theta$ V2.

UniaxialRot.exe uses the principles expounded above to transform a lattice of points into their rotation about an axis running through a point. It is a C# program developed in C# Microsoft Visual Studio.net 2003 environment. The source program and resources are available in the direct-ory Accessory/Transform.

When we run the shortcut for "Mirror.exe" found in the Ac-cessory/Transform directory, we create a dialogue box with input fields "Input File", "Output", "point on mirror", "direc-tion vector 1" and "direction vector 2".

The Input File field is for the name of a valid text file that contains the list of lattice points making up our 3 dimen-sional object. We note that as a default we have "..\..\in-put.txt". The '..\..\' is recommended since this relates to the \Transform directory containing data files (the '.exe file runs from '.\bin\release' directory).

The Output File field is for the input of an output file name. If this does not previously exist it will be created in opera-tion. Again it is recommended to retain the prefix '..\..\' to make sure we create it in the \Transform directory.

In example, to demonstrate the functioning of this pro-gram, Arrow.txt is a file a lattice of points describing a simple arrow lying parallel to the x axis at a z level of –1 between X=+2.5 and X=+3.5. To reflect this object in the plane representing the z=0 plane, we need to choose an arbitrary point lying in the plane eg. (0,3,0) and 2 vectors

parallel to the plane e.g. $\begin{pmatrix} 3 \\ 0 \\ 0 \end{pmatrix}$ and $\begin{pmatrix} 1 \\ 2 \\ 0 \end{pmatrix}$.

These values may be used for our input fields. Note if the input value is not a valid number an error message is displayed, just press continue and correct the error.

Set the Input File field to "..\..\Arrow.txt". Now to run the transformation we must press the 'Open' button. Go to the \Transform directory and check for the name of the output file (default MirrOut.txt) the details should show that it has been freshly created. Open it with "Notepad". On examination, you may see the first two lines describe the reflection, then the first three columns describe the initial arrow and the next three describe the transformed arrow. The last 5 lines describe the plane of the mirror and point on the mirror.

Use edit->select all and edit->copy then close the file. Open Transform.xls from the Transform directory, select page 'arrow', click cell D8 select edit ->paste. The XY, XZ, YZ charts illustrate schematically the initial and transformed arrow. Changing values and press the 'Open' button will cause a new transformation and the appropriate cut and paste operation to cell D8 on 'arrow' page of Transform.xls from the newly formed output file will show the visualisation.

When you are finished you can terminate the application by clicking the 'Close' button.

Consider a 90° rotation of Arrow about an axis lying at 45° in the XY plane and passes through the origin.

To do this, click the shortcut to UniaxialRot.exe. This opens the UniaxialRot dialogue box, which has input fields for Input File, Output File, point of rotation, axial direction vector and angle of rotation.

We may leave the point of rotation as the default (0.0, 0.0, 0.0), change the axial direction vector to $\begin{pmatrix} 4.0 \\ 4.0 \\ 0 \end{pmatrix}$ and set angle of rotation to 45. Change the field of Input File to "..\..\arrow.txt" then press "Open" button. Go to Transform directory and check the details of the output file. They should show it to be freshly created. Open it with notepad.

The first 2 lines describe the transformation. The next 8 lines represent arrow. The first 3 columns describe its initial position and columns 5,6,7 describe its transformed position. The next 15 lines describe Cartesian axis and its transformation. The last three lines describe the rotational axis.

Choose edit ->select all, edit ->copy and close the file. We go back to Transform.xls and 'arrow' page select D8 then select edit ->paste. The charts XY, XZ, YZ now illustrate the schematic of original and transformed arrow and Cartesian axes.

You can change the transformation by changing the parameters on the UniaxialRot dialogue box and pressing the 'Open' button, open the new output file and use edit ->select all, copy and paste to the appropriate place in Transform.xls to realize the visualisation.

The input file has a structure, which consists of a header of 3 lines. These lines are not used in computation. Each of the following lines represent a point in the latticework x,y,z values each separated by spaces or tabs and terminated by 'return'. This format is easily duplicated for any new data set representing a new object.

You may find it interesting to experiment with transformations to 'dart.txt' and 'house.txt' and you can find their visualisations on pages 'dart' and 'house' in Transform.xls. House4.txt is a composite input file that demonstrates how we create complex models by the serial addition of transformed coordinate sets. So to create House4.txt, from house.txt a second neighbouring house was generated by

52

reflecting in the ZY plane at x=-40.4. The two resulting houses were then reflected in a plane lying predominantly in XZ at Y=-75, the z direction vector slightly offset in +x direction. This represents houses on the opposite side of the street lying on a mild gradient. To demonstrate how we can replicate this block 4 houses into a street, if we rotate about point (40.4, 1460, 0.0) with a vertical axial vector $\begin{pmatrix} 0.0 \\ 0.0 \\ 5.0 \end{pmatrix}$ and an angle of 12.04° that ensures the two plots do not intersect.. Run UniaxialRot.exe with House4 and the above parameters. If we copy and paste the resulting output file content to Transpose.xls page "MultiHouse". Use copy and paste for each representative house. On this page houses 1&2 are represnted by block M105 to R187, houses 3&4 U105 to AA187, houses 5&6 AD105 to AJ187 and houses 7&8 AM105 to AS187. The result of such a transformation is displayed on charts of Transform.xls page "MultiHouse".

It is easily seen that through replication and appropriate transforms we can produce a street and then a complete estate. As such we have devised a very powerful modelling tool. Thus these routines can be the major functions in a software engine for creating landscapes. With the addition of perspective and rendering we may be able to calculate views and create tours through virtual scenes. These transforms may also be the core functions in software designed for coordinates conversion between alternate systems.

Chapter 4

IMPEDANCE MATCHING

Fig. 11

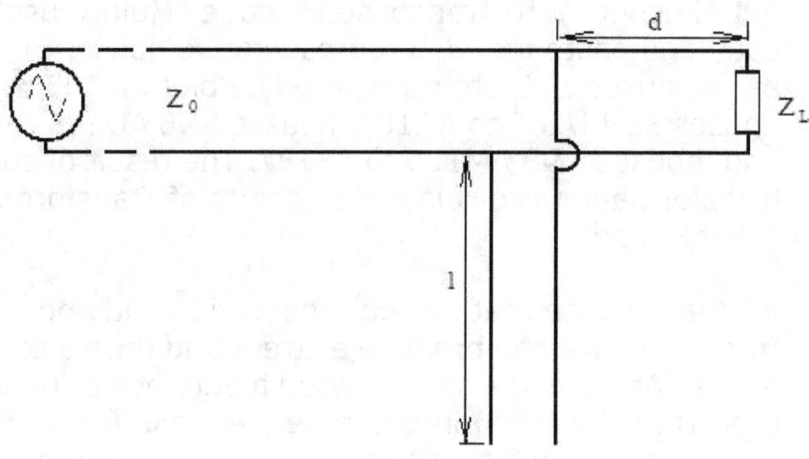

Phase matching for transmission lines is the taking into account the wave superposition properties of elements of the circuit. These become significant when the wavelength is large relative to the characteristic sizes of components. That is, we are taking into account wave reflections from discontinuities in the circuit.

Lossless lines are important and frequently used as transmission line models, they idealise the transmission of the signal by ignoring Ohmic losses. These may represent coaxial cables, twisted pairs or microwave channels. In onboard circuitry they represent onboard signal carrying tracks.

Lossless lines are considered composed of a distributed length of serially connected inductors regularly short-circuited by shunt capacitors as illustrated in figure 12.

The properties of the line are characterised by a serial inductance and shunt capacitance per unit length. This is represented by $Z_0 = \sqrt{L/C}$, a real quantity, where L is inductance per unit length and C is the capacitance per unit length. The phase velocity in the line is $\frac{1}{\sqrt{LC}}$.

Fig. 12

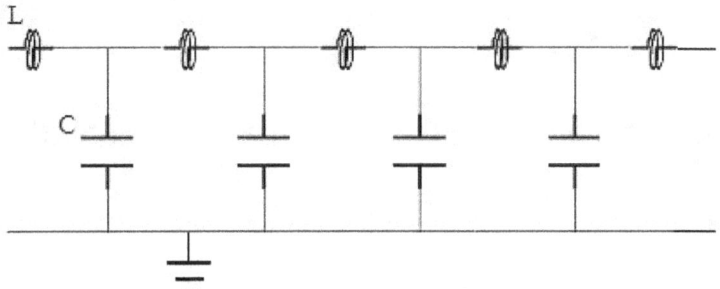

We use complex notation and arithmetic to handle sinusoidal signals. Thus Ohm's law becomes Z= V/I where Z is a complex impedance and V and I are also complex. Without derivation we can state the voltage reflection coefficient Γ , the ratio of amplitude of the propagated wave with the reflected wave (see figure 13), thus

$$\Gamma = \frac{V^-}{V^+} = \frac{Z_L - Z_0}{Z_L + Z_0}$$ *Eqn.* **4.1**

Fig. 13

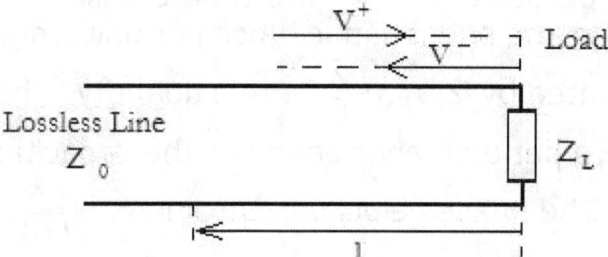

The average power transmitted to the load on the matching of Z_0 Z_0 to the real part of load impedance Z_L resolves to :-

$$P_{avg} = \frac{1}{2}\frac{|V_0^+|^2}{Z_0}\left(1-|\Gamma|^2\right) Eqn.\ 4.2$$

The ratio of Vmax to Vmin is called the standing wave ratio (SWR).

$$SWR = \frac{V\max}{V\min} = \frac{1+|\Gamma|}{1-|\Gamma|} Eqn\ 4.3$$

The power that is not realised in the load is reflected down the line as the return loss (RL). This is defined in decibels.

$$RL = -20\log|\Gamma| Eqn\ 4.4$$

As we travel back down the line away from the load, the interaction of the forward wave and the return wave cause the local detected signal, voltage, current and impedance to vary. This is the observed as a variation in SWR and reflection coefficient and depends on the wavelength λ at the measured frequency.

Let $b=2\pi/\lambda$, at length l away from the load

phase angle $\qquad \varphi = \frac{2\pi\ l}{\lambda} \quad Eqn\ 4.5$

The impedance looking into Z_L from length l away from Z_L is given :-

$$Zin = Z_0 \frac{Z_L + jZ_0 \tan\varphi}{Z_0 + jZ_L \tan\varphi} \quad Eqn\ 4.6$$

Equations 4.1,4.2,4.3,4.4,4.5 and 4.6 are fundamental equations that are used in the solution of transmission line problems.

A quarter-wave transformer is a common device in load matching. In this device a length of transmission line equivalent to $\frac{1}{4}$ the wavelength at the required frequency is interposed between the load and the main transmission line.

Fig. 14

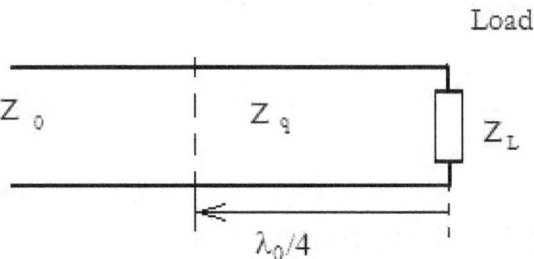

Z_0 characteristic impedance of main transmission line.
$\quad Z_q$ characteristic impedance of quarter-wave transmission line.

Z_L load impedance.

Equation 4.6 when $\phi = \pi/2$, i.e. when $l = \lambda/4$, $\tan \phi$ explodes and the equation becomes

$$Zin = Z_q \cdot \frac{Z_q}{Z_L} = \frac{Z_q^2}{Z_L} \quad Eqn.\,4.7$$

If $Z_q = \sqrt{Z_0 Z_L}$ simplifies to $Zin = Z_0$ and $\Gamma in = 0$.

To see the filter characteristics of a quarter-wave transformer, we have to consider the value of equation 4.6 as frequency varies through f_0 .

Example 10

A 100 ohm load line is loaded with lumped elements, a load impedance 120 + 70i ohms. This is converted into a real impedance by a shunt. Find the value of the shunt impedance and the characteristic impedance of the quarter--wave line used to match the load.

a) Lumped Element Solution

$Z_0 = 100\,\Omega\, Z_L = 120 + 70j$

normalised impedance (Z/Z_0)

$z_0 = 1.0\, z_L = 1.2 + 0.7j$

convert to admittance

$y_0 = 1.0\, y_L = \dfrac{1.2 - 0.7j}{1.93}$

then set $y_S = \dfrac{+0.7j}{1.93} = 0.3627\,j$

58

$$y_L + y_S = \frac{1.20}{1.93} = 0.6218, \frac{1}{y_L + y_S} = 1.608 = zin$$

$Zin = 160.8 \, \Omega$

Given $y_S = 0.3627 \, j, z_S = -3.0649 \, j, Z_S = -306.5j \, \Omega$.

b) Using a Smith Chart
1. Plot normalised $Z_L = 1.2 + 0.7j$
2. Reflect it through $1.0 + 0.0j$
3. Use the constant y_{real} circle to work out y_S and new operating point
4. Reflect through point $1.0 + 0.0j$ to read Zin

Fig. 15

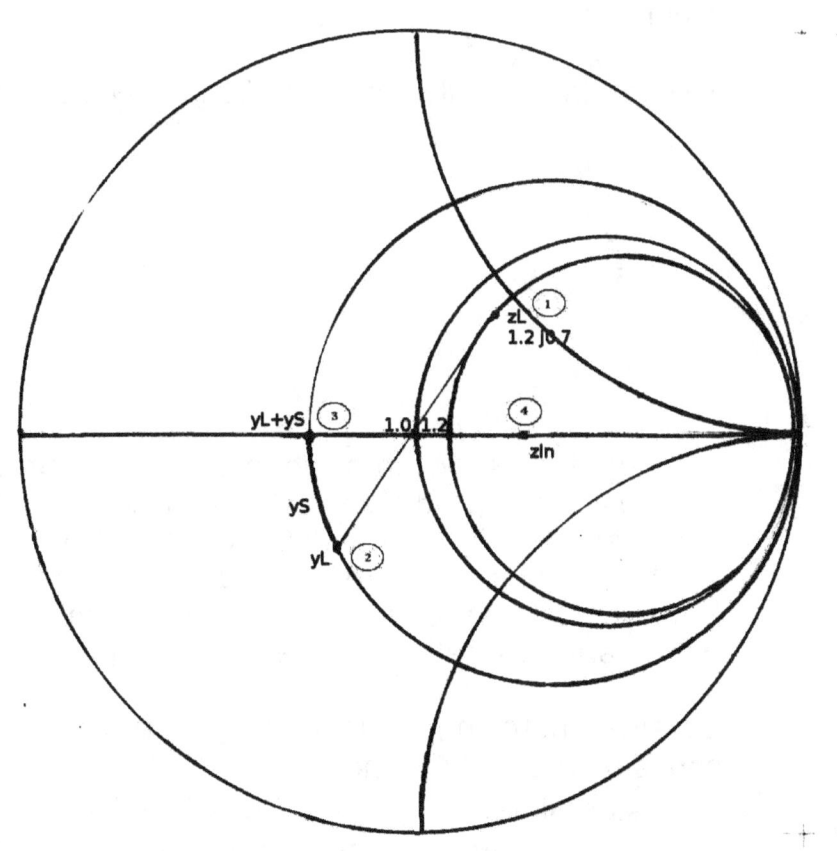

Thus the lumped element combination of Z_L shunted with Z_S will give a resultant 160.8 + 0.0jΩ.

Zin can be matched to the 100Ω transmission line using a quarter-wavelength line. The required impedance of this quarter-wave line is given:

$$Z_q = \sqrt{Z_0\ Zin} = \sqrt{100 \times 160.8} = 126.8\Omega$$

Quarterwave.exe found in directory Accessory/Transm uses text file QWVIn.txt to input values of Z_0, Z_L the operating frequency f_0 and the frequency range increment (as a ratio of f_0) i_0. It is used to examine the properties of our filter. Quarterwave.exe calculates the ideal impedance of the quarter-wave transformer line and then calculates the input impedance and reflectance over the range of frequencies around f_0.

To examine our filter around 100 Mhz we use input::

```
100.0   0
160.8
1E8
0.5     1.5
```

0.005

Save the file as QWVIn.txt then execute Qaurterwave.exe. The results are produced in QWVOut.txt. This file can be imported to QWV.xls. Close QWVOut.txt and open QWV.xls on sheet1 select F17, from menu select Data->Refresh Data and select the output file QWVOut.txt. The result is a plot of absolute reflectance against frequency.

Consider matching the load using a stub line placed in series on the Z_0 Z_0 line.
This is possible since $Z_L = 1.20 + 0.7j$ lies within the $1+ jx$ 1 + jX circle on the Smith Chart. The two important parameters to be determined are d, the distance from the load to the stub line and l, the length of the stub line.

60

On the Smith Chart we plot the normalised load $1.2+0.7j$. It lies within the $1+jx$ as noted above. If we draw the SWR circle that passes through it centred at $1.0+0.0j$ we note as we follow it round in a clockwise direction (inductance towards the generator), the two points at which it cuts the $1+jx$ circle. P1 is 1 – 0.75j, P2 is 1 + 0.75. Checking the points on the scale 'wavelength towards the generator' we have

$$Z_L = 0.1739\lambda, \ P1 = 0.3495\lambda, \wp P2 = 0.1517\lambda \ \overline{Z}$$

Thus
$$Z_L \ to \ P1 = 0.1756\lambda, \qquad Z_L \ to \ P2 = (0.5 - 0.1739) + 0.1517) = 0.4778\lambda$$
away from the load. The position d of the stub line is 0.1756λ or 0.4778λ away from the load or any of these values plus n×0.5λ, where n is an integer. The preferred distances are the lowest to get the best filter and to minimize dispersive effects.

At these points P1 = 1 – 0.75j and P2 = 1 + 0.75j. At P1 we must add +0.75j to centre the operating point and for P2 we must add –0.75j to centre the operating point. This is all shown in figure 16.

Fig. 16

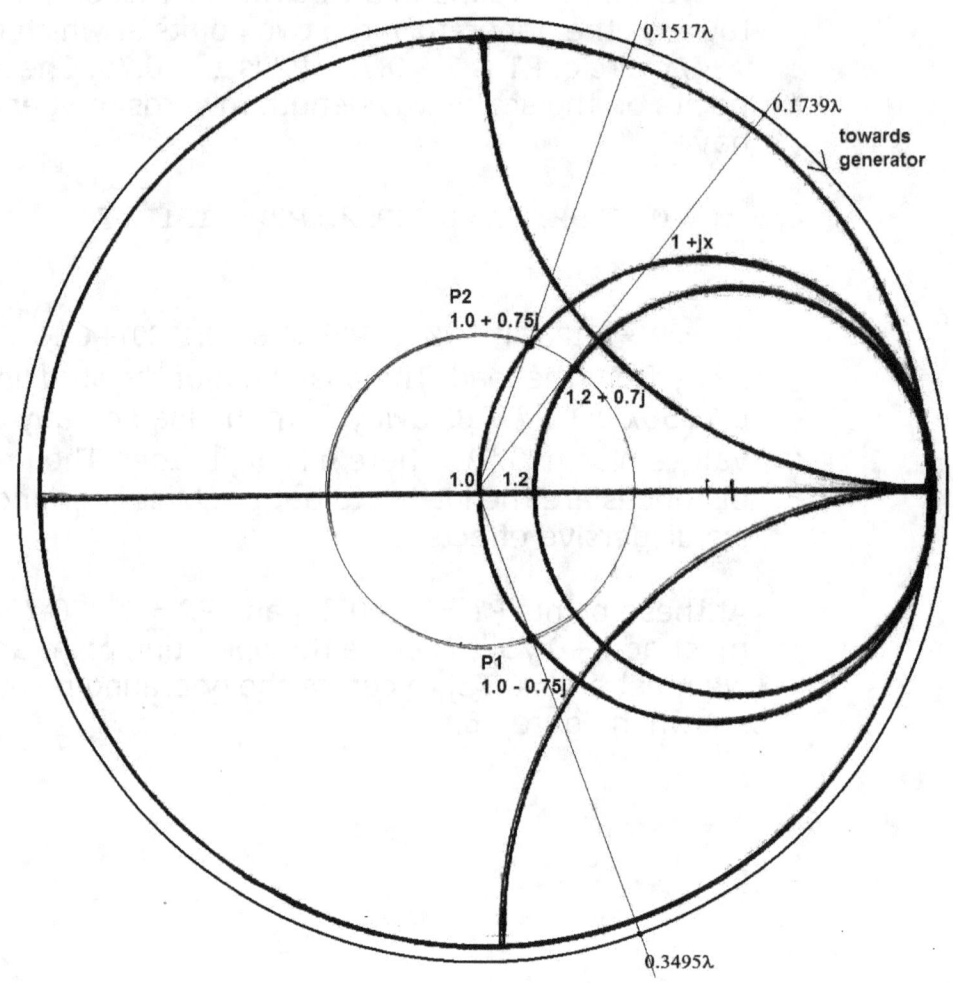

The graphical representation using the Smith chart is limited in accuracy according to the scales on the chart and the tolerance of our plotting. Zintline.exe found in folder Accessory/Transm is program that calculates Zin over different distances from the load on the input transmission line.

Open ZintlineIn.txt and enter :-

100.0 0.0
120.0 70.0
0.0   1.0   0.01

Representing:-
$Z_0 = 100.0 + 0.0\,j\Omega, \qquad Z_L = 120.0 + 70.0\,j\Omega$
and d from $0.0\lambda\ to\ 1.0\lambda$ in increments of 0.01

We save it and run Zintline.exe. ZintlineOut.txt gives results in d/λ written as l/λ. Looking in column 3 for Zin real we see we expect Zin real to be 100.0 at about l/λ =0.18 and 0.48. We can interpolate to obtain a more accurate answer of Zin real =100.0
$l/\lambda = 0.1779$, Zin =100.0 - 66.36j
$l/\lambda = 0.4789$, Zin =100.0 +66.53j
This more accurate answer is in good agreement with those estimated using the Smith chart, but the imaginary values are significantly different, ±66.5jΩ as compared with ±75jΩ.

It is now left to determine the length of stub to be used. We have two files OpenCircOut.txt and ShortCircOut.txt. If we assume an open circuit stub line with characteristic impedance Z_0 we may use Open CircOut.txt or if we use a similar stub short circuited then we may use ShortCircOut.txt.

With an open circuit stub we find we get a relative impedance close to - 0.665j at $l/\lambda = 0.16$, and close to+0.665j $l/\lambda = 0.34$. With a short circuit stub we obtain a relative impedance close +0.665j at close to l/λ = 0.09 and relative impedance close to –0.665 with $l/\lambda = 0.41$. The actual values can be estimated by interpolation.

Thus the solution of load matching using stubs will appear as below

Fig. 17

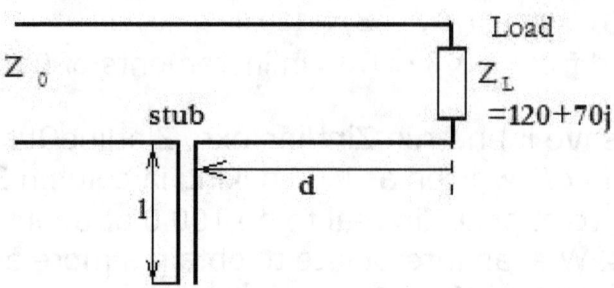

d	l open circuit stub or	l closed circuit stub
0.1779 λ	0.3433 λ	0.0933 λ
0.4789 λ	0.1567 λ	0.4067 λ

Summary

In this book we have had a walkthrough of a number of examples demonstrating mathematical techniques with direct practical application. We use applied techniques of matrix applications, vector analysis, complex number arithmetic and integration of trigonometric functions. We have created tables and used spreadsheets and shown how a correctly designed spreadsheet can give a visualisation of a process and have shown how we may use data exchange techniques to give easy updates.

I have limited the complexity of problems to give "tasters" to the possibilities and strength of the tools we are using. In linear programming, spreadsheets created for our solutions are flexible and accurate, showing how these tools can be used to model processes and scenarios on a purely theoretical level predicting very potent outcomes.

LinEqSoln.xls contains a spreadsheet tool I have created for solving linear equations. It has been used to determine unit costs in Examples 1 and 2. In Examples 3 and 4 the same tool is used to show how linear equations are used in solving dc current circuits, in examples 4 and 5 we are introduced to routines that can solve linear equations in rational and complex numbers that we use to investigate AC circuits. In Example 6 I have shown how a simple spreadsheet can model production scenarios and give immediate and easily understood results, facilitating the optimisation of resource usage. Example 7 is a classical "Cost of Supply" transportation problem that models the optimisation of supplying different demand points at minimum cost.

In Examples 8 and 9 we consider the trajectory of an object in still air. We produce a solution that can be adapted for different objects in different fluids, given the relevant properties. We see as expected, at low velocity the deviation from the frictionless model is mild but this is very important at initial speeds of order of more than 100m/s. In the section 3 we create and demonstrate the use of net drawing and transformation typically used at the core of modern graphics animation software.

In the final section we look at electrical engineering and phase matching in RF circuits.

As explained, the routines and programmes created and used in this project are all generic and will run under a standard "Windows" 32bit platform and the code and re-sources have been made available in the directory and sub directories of "Accessory" found as downloads from my website at www.anglo-african.com. Possibly the easiest op-tion is to download the zip file Accessory.zip, that contains the full file structure.

BIBLIOGRAPHY

M. S. Makower and E Williamson.
Operational Research, 4 ed.
Teach Yourself Books, Hodder and Stoughton, 1985.

Benjamin Crowell,
Electricity & Magnetism
www.lightandmatter.com

C.P. Kothanandaraman, S Subramanyan
Heat and Mass Transfer Data Book, 3rd ed.
Wiley Eastern Ltd., 1978

R.P. Gillespie
Integration
6th ed.
Oliver and Boyd, 1963

Cristopher Clapham and James Nicholson
the Concise Oxford Dictionary of Mathematics
4th ed.
Oxford Reference, 2009

G. M. Phillips (Author), P. J. Taylor. **Theory and Applications of Numerical Analysis**, 2nd ed. Academic Press, 1996.

B.I. Bleaney And B. Bleaney
Electricity and Magnetism,
2nd ed., 1965

J.R. Calvert and R.A Farrar
An Engineering Data Book
Palgrave, 1999

David M Pozar
Microwave and RF Design of Wireless Systems
John Wiley, 2000